Quail Plantations
of South Georgia and North Florida

Quail Plantations
of South Georgia and North Florida

PHOTOGRAPHS BY HANK MARGESON

TEXT BY JOSEPH KITCHENS

The University of Georgia Press   Athens & London

© 1991 by Hank Margeson
Published by the University of Georgia Press
Athens, Georgia 30602
All rights reserved

Set in by Simoncini Garamond by G&S Typesetters, Inc.
Printed and bound by Thomson-Shore, Inc.
The paper in this book meets the guidelines
for permanence and durability of the Committee on
Production Guidelines for Book Longevity
of the Council on Library Resources.

Printed in the United States of America
95 94 93 92   C   5 4 3 2

Library of Congress Cataloging in Publication Data

Margeson, Hank.
Quail plantations of south Georgia and north Florida /
photographs by Hank Margeson ; text by Joseph Kitchens.
p.   cm.
ISBN 0-8203-1386-6 (alk. paper)
1. Quail shooting—Georgia. 2. Quail shooting—Florida.
3. Plantations—Georgia. 4. Plantations—Florida. I. Kitchens,
Joseph. II. Title.
SK325.Q2M37 1991
799.2′48617—dc20   91-3054
                                                      CIP

British Library Cataloging in Publication Data available

Frontispiece: Pebble Hill Plantation,
Thomas County, Georgia

*To Laura and Carrie Jane, Bob,
Carolyn, Keith, Jane, and Charlie Mac*

Contents

Preface ix

Acknowledgments xi

Plantation Country, Past and Present 1

Gracious Living 7

Plantation Portraits 33

Southern Landscapes 53

Quail Shooting 77

Preface

WELL OVER a century ago, quail shooting plantations replaced the cotton growing plantations in south Georgia and the Florida panhandle. Although this more sporting life-style has long since flourished, a certain mystique endures about the plantation traditions of old. By approaching the contemporary plantation culture in a documentary context it was my intention to discover and then portray an objective reality. But a photographic documentary is inherently subjective, and the content of this book was shaped by the many influences in my life and career: the hunting tradition in my family, the solace I find in the changing yet unchanging landscape, a familiarity with the social order and nature of the people, an interest in the customs that give character and identity to the region, my training in art, my teaching philosophies, and a preference for certain visual styles. An accurate portrayal, then, actually becomes an honest attempt at visual integrity but with an acknowledged bias—a point of view.

The focus of my efforts was on the people who work on the plantations, the structures, the environment, and the activities surrounding the hunting event. I discovered that the present was physically and emotionally intertwined with the past.

The scope of this project proved much greater than I had originally expected. What began in the summer of 1985 as a one-year project became an endeavor that continued through the fall of 1990, resulting in thousands of images, few of which would have been possible without the help of those friends who share my interest in documentary photography and the cultural heritage of the region. This book is a testament to the many unselfish individuals and organizations whose efforts made its publication possible.

Initial funding for the project was provided by an Artist Initiated Grant from the Georgia Council for the Arts. Thomas Jones, former curator at the Albany Museum of Art, sponsored my application for the grant, exhibited the resulting work, and arranged a series of traveling exhibitions. Malcolm Call, director of the University of Georgia Press, provided the framework that enabled me to turn the collection of photographs into a book. John Head, of the University of Georgia, assisted in making order and cohesion from the hundreds of prints considered for publication. Bob Owens at North Georgia College afforded me the time to refine the images while balancing these efforts with my teaching duties.

I spent five years in the field working with my Crown Graphic and Nikons, and the work has greatly benefitted from the contributions made by family and friends. Mike Murphy introduced me to his numerous friends on the plantations of Georgia and Florida; Robert Margeson encouraged my efforts throughout the project with his rich understanding and appreciation for quail hunting and southern heritage. Joe Kitchens provided valuable insight into the historical setting as well as contemporary conditions on the plantations. Heywood Parrish, David Campbell, Edgar Campbell, Sally Sullivan, George Moreland, Hilly Thompson, Chip Hall, and other plantation owners and managers welcomed me into their private worlds and enabled me to discover what lay beyond those veils of Spanish moss, the pristine homes, and fields of tall pines. I owe them all a special debt. Thanks also to Patsy Martin, Jane Hume, Ken Garey, Joe Cronan, Doug Brown, Bob Nix, Jane Jackson, Burke Walters, Gator Lee, Greg Oyer, and the many others who helped.

Above all, it was the warm and tireless support of my family that allowed me to pursue the project to this end. For the valued part they played in this production I am deeply grateful.

HANK MARGESON

Acknowledgments

MANY INDIVIDUALS AND ORGANIZATIONS have supported this project by acquiring photographs from the plantation portfolios for their collections. Others have generously contributed toward the cost of producing the body of work from which the images in this book have been selected. It is with grateful appreciation that I recognize these collectors and patrons who helped bring the idea to this present form.

Blockhouse Restaurant, Greenville, S.C.
Campbell Farms, Albany, Ga.
Georgia Council for the Arts, Atlanta, Ga.
Mr. Andrew Lewis Ghertner, Atlanta, Ga.
Mr. and Mrs. J. Keith Skelley, Roswell, Ga.
Storer Cable Communications, Albany, Ga.
Win-Tex, Dallas, Texas

Citibank, New York, N.Y.
Mr. and Mrs. R. M. Margeson III, Albany Ga.
Mr. and Mrs. R. H. McGarity Sr., Albany, Ga.
Merry Acres Development Co., Albany, Ga.
Plantation Services, Albany, Ga.
Mr. and Mrs. John M. Simmons, Bainbridge, Ga.

Albany Museum of Art Board of Trustees, Albany, Ga.
Alltel Mobile, Albany, Ga.
Carlton Company, Albany, Ga.
Mr. and Mrs. John J. Carney, Albany, Ga.

Chattahoochee Valley Art Association, LaGrange, Ga.
Citizens and Southern National Bank, Tifton, Ga.
Columbus Museum, Columbus, Ga.
Coney Lake Lodge, Leesburg, Ga.
Mr. Patrick D. Dugan, Albany, Ga.
Ms. Julia Evatt, Ellijay, Ga.
High Museum, Atlanta, Ga.
King & Spalding, Atlanta, Ga.
Dr. and Mrs. W. Thomas Mitcham, Albany, Ga.
Peterson Young Self & Asselin, Atlanta, Ga.
Mr. and Mrs. Thomas O. Powell, Atlanta, Ga.
Reed Brothers Hardware, Albany, Ga.
Mr. and Mrs. George G. Riles, Albany, Ga.
Mr. and Mrs. Douglas E. Smith, Gainesville, Ga.
Mr. and Mrs. Denis G. Toomey, Kilkenny, Ireland
Mr. and Mrs. G. Lee Underwood, Atlanta, Ga.
Dr. and Mrs. J. David Whittle, Albany, Ga.

Quail Plantations
of South Georgia and North Florida

Plantation Country, Past and Present

BEYOND THE TIDEWATER of the southern colonies in England's American empire, there was once a vast desert of wire grass and pine barrens. It followed the Piedmont Plateau in a widening ribbon that stretched southward from New Jersey to Georgia. There was little here to entice settlers into this coastal plain. Its broad expanse was largely ignored by the flow of immigrants lured by grants of free land in the emerging cotton belt above the shoals of the Savannah, Ogeechee, Altamaha, and Chattahoochee rivers. Settlement remained sparse well into the nineteenth century.

While most of the southeastern wire grass region was seen as inhospitable and unprofitable, an area north of Tallahassee, that included the Tallahassee Hills and the broad basin of the Flint River emerged as a prosperous agricultural region in the decades before the Civil War. Its diversified agriculture, heavier soils, and abundant water made the region attractive for investment, settlers, and the slave system. In particular, the soil around Albany and Thomasville in south Georgia, and Tallahassee in north Florida, was uniquely suited for cotton growing. A large plantation culture flourished around these three cities. By the late 1850s, the area had become a major food-producing region upon which the Confederacy would rely heavily during the war years.

Plantations continued to predominate even after emancipation ended the slave economy upon which the plantation system was based. The rich land remained profitable, and the social order continued functionally intact because of the area's geographic isolation and the apparent lack of economic alternatives available to the freedmen. Not until the nationwide agricultural depression of the late 1870s did the system break apart. But unlike other areas, the Albany-Thomasville-Tallahassee region began to experience an influx of money and, in a limited way, prosperity. Fueled by industrial capitalism, newly affluent northerners began to find the area appealing for investment and winter recreation. Accessible by rail from the great urban centers of the north, it became a popular winter retreat, a haven from the snow and freezing temperatures of New England and the Old Northwest.

Thomasville boasted seven large hotels in the 1880s and was ringed by a carriage boulevard where visitors from the north enjoyed what was then considered to be healthful, healing air generated by giant long-leaf pines. While this era was short-lived, it did establish the region as a sportsman's winter paradise, a legacy that led directly to the present-day shooting plantation.

Several forces were at work to encourage the emergence of a kingdom of shooting preserves in southwest Georgia. The availability of large tracts of land at low prices, refreshingly mild winters, cheap labor, and the traditional southern hospitality made the purchase of a plantation attractive and affordable.

The industrialization of the country's northern cities fostered a new class of wealthy entrepreneurs. They were often immigrants, if not to the United States, at least to the new and rapidly growing cities of the heartland. Proud and successful, these newcomers were eager to participate in class traditions similar to those of the Europe that had spawned their Irish, Scottish, English, or other European ancestors. Perhaps the inspiration was only mild winters and great shooting, but one is tempted to think that a new American aristocracy was confirming its social station by donning the mantle of the earlier southern gentry.

In addition, an American style of wing shooting had emerged that focused on game birds whose protective instinct was to remain still, at least briefly, before the pointing and flushing dogs. Shoots on English estates more typically involved driven game, especially grouse and pheasant. In southwest Georgia and north Florida, the most plentiful game birds were dove and quail. The former were primarily migratory and were shot on the wing over the fields that had been harvested or even left to entice the birds. Quail, however, were plentiful year round and were exciting quarry because of their habit of forming large coveys that tended to remain in a still cluster when the dogs approached. Scenting the birds, the dogs pointed or flushed according to their breed and training.

Ralph Thrower, Horseshoer
Mercer Mill Plantation
Worth County, Georgia

Exploding in a flurry of feathered blurs, quail made a challenging and elusive target.

The century that dawned in 1900 also witnessed the beginning of what is known as the Golden Age of Shotgunning. In England and the United States, the shotgun was refined to perfection. Although breach loading and smokeless powder were the most obvious innovations, subtle improvements were made in boring, choking, engraving, and checkering, and the guns became works of art admired and written about by succeeding generations of shooters.

In their traditional form, shotguns with side-by-side double barrels are still the preferred guns on the plantations. The smaller bores, 20 and 28 gauge, are thought to be more sporting than the heavier 12s and 16s. In an age when most shotguns are either autoloaders or pumps that can fire three or more shots, the modern mechanical guns are sometimes considered unsporting on quail plantations. Gun talk is likely to run to Parkers and Purdeys, L. C. Smiths and Winchesters, or the other makers of fine double-barreled shotguns from generations past.

Although many other types of game were plentiful, quail shooting became the favored sport. Perhaps this was because it occurred during the fall, a season of traditional holiday festivities and reunions. Or perhaps it was the spectacular autumn weather, with its crisp November skies and golden-brown terrain; or that indescribable sensation of a covey rise over a pair of rock-steady pointers. While not all plantation owners came for the shooting, most did enjoy the social aspects of the event: leisure rides in the hunting wagons, bird-dog field trials, catered lunches in the field, and, of course, the pleasure of visiting with family and houseguests.

The land-grant system provided the framework for settlement, but it was often men and women of some affluence and skill from more settled areas who made the land productive. The importation of slaves from Africa had ceased before the area was opened, so even the work force of the new plantations came from settled cultures, bringing with them the traditions and social structure of the piedmont cotton and tidewater rice plantations. Southwest Georgia, like any other new frontier, saw a rich merging of the earlier experiences and customs of its newly arrived residents, both black and white. At the same time, because the area was somewhat isolated in the years immediately following the Civil War, it retained many vestiges of antebellum plantation life that were rapidly vanishing elsewhere. It was this way of life that the new plantation owners sought to emulate and the old ones wished to preserve.

Although debatable, it was a common assumption in the late nineteenth century that the southern planters were the only true American aristocracy. Interest in southern plantation culture was no doubt sparked by the publicity accompanying the novel and movie "Gone with the Wind" in the 1930s, which tempted the new plantation owners to see themselves as the conservators of this proud tradition. Indeed, local lore has it that the first showing of the movie outside Hollywood was to a gathering of plantation owners at Melrose Plantation, near Thomasville.

The new hunting plantations and game preserves were modeled after the old cotton plantations, which today's terminology reflects. Staff housing is known as the "quarters," managers are often referred to as "overseers," the owner's residence is the "main house" or "big house." From the postbellum era came the inspiration for the "plantation stores," where the freedman once bought goods on credit from his former master and which now occasionally survive as general stores for the workers and their families who are far removed from the convenience of town.

The word *plantation* usually conjures images of magnolia trees, live oak alleys, and stately Greek Revival mansions. Certainly all these are to be found in abundance. But fire has destroyed many of the classic southern structures, and although there are traditional residences like Dixie, Susina, and Pinebloom, others vary in architectural style from Tudor fantasies to rustic lodges.

The homes, the land, the people, and the culture they created owe a great debt to the nurtured and plentiful presence of bobwhite quail. Known as "the quail capital of the world," this region today contains more than a hundred working plantations, many covering thousands of acres each. The quail hunting plantations that blanket south Georgia and north Florida like a patchwork quilt have proved more durable than the cotton kingdom they replaced. Many symbols of this century-old culture are preserved in this collection of photographs by Hank Margeson. His visual documentary has captured the character of twentieth-century plantation life. These images are evidence that some of the south's most colorful traditions still survive.

Gracious Living

THE CLASSICAL ARCHITECTURE of the antebellum plantations suggests a gentry rooted in the land. In the areas opened to settlement in the 1820s and 1830s, this agrarian culture and its slave-based economy were destroyed within a generation, victims of civil war and international economic revolution.

Ironically, for nearly a century many antebellum houses and plantations have been preserved by northern families as winter retreats, especially in the Red Clay Hills north of Tallahassee and in the Albany area. Those that have burned or been otherwise destroyed have in many cases been replaced by neoclassical houses. Columned porticos convey the gracious style associated with the gentility and social system of the old southern aristocracy. In fact, many are owned by the heirs of this aristocracy, or by families that have shared in the prosperity of the modern south. The traditional interiors of the grand homes are hospitable. Paintings, prints, and sculptures reflect a love of nature and the hunting traditions of the area. Spiral staircases, gracious dining rooms, and cabinets filled with Smiths, Parkers, and other fine guns evoke the privileged style of plantation life. Old and new masters of animal art often adorn stately rooms, as do photographs of field trial champions and portraits of prize horses. Family and guests gathered around open fireplaces set the mood of the shooting season and rekindle the traditions and romance of bygone days.

Always prevalent are stables and barns, hunt wagons and mules, and the aroma of saddle leather and hay. Kennels hold the promise of tomorrow's hunt. Nearby "quarters" or the "plantation store" nearby remind guests of the many hands required to maintain and preserve a plantation community. The hunting plantation is classic and proud, earthy and wedded to nature.

Entrance Gate
Boxhall Plantation
Thomas County, Georgia

Seminole Plantation
Thomas County, Georgia

Cypress Pond Plantation
Dougherty County, Georgia

Springwood Plantation
Grady County, Georgia

Muckalee Plantation
Lee County, Georgia

Southern Heritage Plantation
Dougherty County, Georgia

Chinquapin Plantation
Thomas County, Georgia

Susina Plantation
Thomas County, Georgia

Southern Heritage Plantation
Dougherty County, Georgia

Dixie Plantation
Jefferson County, Florida

Chinquapin Plantation
Thomas County, Georgia

Entrance
Gillionville Plantation
Dougherty County, Georgia

Oakland Plantation
Dougherty County, Georgia

Pinebloom Plantation
Baker County, Georgia

Loggia, East Wing
Pebble Hill Plantation
Thomas County, Georgia

Drawing Room
Pebble Hill Plantation
Thomas County, Georgia

Dining Room
Gillionville Plantation
Dougherty County, Georgia

Stairs
Gillionville Plantation
Dougherty County, Georgia

Horse Stable
Mercer Mill Plantation
Worth County, Georgia

Horse Stable
Pineland Plantation
Baker County, Georgia

Cow Barn
Pebble Hill Plantation
Thomas County, Georgia

Hunting Wagon and Stable
Kelly Pond Plantation
Jefferson County, Florida

Horse Stable
Hogan Place (formerly Melrose Plantation)
Thomas County, Georgia

Ichauway Store
Ichauway Plantation
Baker County, Georgia

Plantation Portraits

PLANTATIONS ARE COMMUNITIES of people who live, work, and rear their children together. Gardeners keep the grounds, and housekeepers see to the cleaning. Farm hands harrow miles of firebreaks for the annual burning and plant the feed patches of corn and Egyptian wheat for quail and dove. Stock must be fed, stables and kennels cleaned, dogs and horses tended, trained, and handled. Here men and women nurture the land and animals, and in turn their own nature is formed.

For both owner and help there is pride in training a champion pointer and pleasure in the spirited or gentled nature of the horses. A boy's first gun or dog, or a handler's succession to trainer are times to be remembered. Victory or failure in the field trials is shared by owners, trainers, and kennel hands alike. Children who wait for school buses and ride their bikes on rutted or sandy roads know the smell of burning underbrush or freshly broken ground. For the families who live year-round on the plantations, the years are punctuated by the annual rituals and daily chores that make up the tapestry of life here.

When the shooting season begins, there is excitement and endless work: birds to be dressed, guns to be cleaned, guests to be fed, and barbecue to be cooked. As the season draws to an end, there is burning and timbering to be done, dogs to be whelped, colts to be delivered, horses to be reshod, grounds to be groomed, and houses to be refurbished. The weather and seasons are felt firsthand and shape the men and women and the land where they live.

Robin Gates with Flatwood Hank
Gates Kennels
Lee County, Georgia

Walker Lee Clinton with Pinehurst Duke
Coney Lake Plantation
Lee County, Georgia

Joe Hicks at Florida National Championship Field Trial
Chinquapin Plantation
Suwannee County, Florida

Courtney Moreland with Buzzsaw's Stormy Bud
Coney Lake Plantation
Lee County, Georgia

Eddie Dennard and Arthur Johnson
Gillionville Plantation
Dougherty County, Georgia

Doc Williams and Richman Margeson
Pineland Plantation
Baker County, Georgia

Men Beside Kennels
Ichauway Plantation
Baker County, Georgia

George Mitchell at Master's Field Trial
Pineland Plantation
Baker County, Georgia

Edgar Campbell
Campbell Farms
Dougherty County, Georgia

John Arthur Butler
Pineland Plantation
Baker County, Georgia

Ralph Thrower
Mercer Mill Plantation
Worth County, Georgia

George Mitchell
Coney Lake Plantation
Lee County, Georgia

Rearing Horse
Pineland Plantation
Baker County, Georgia

Lester Blackshear
Pebble Hill Plantation
Thomas County, Georgia

Family Portrait
Coney Lake Plantation
Lee County, Georgia

Boys on Bicycles
Ichauway Plantation
Baker County, Georgia

Children at Master's Field Trial
South Dougherty Community Center
Dougherty County, Georgia

Chopping Pork Barbecue
Coney Lake Plantation
Lee County, Georgia

Southern Landscapes

SANDY RIDGES of pine, hardwood-lined stream bottoms, slow-flowing rivers, and cypress swamps make up the landscape in plantation country. The land is hilly with red clay soil south of Thomasville and flatter with sandy loam south of Albany. Seasons change with little fanfare. When the gentle rains of February inaugurate the brief days of spring, ground fires are lit to clear fence rows and kill back hardwood encroachment for another year. Gray smoke clouds the forest in the day and fat lighter pine stumps flicker and glow for nights afterward. Azaleas and Cherokee roses flourish briefly in early spring, but by May the first ninety-degree days impose a languorous pace on all life here. Spanish moss that drapes the bare winter trees is lost in a green canopy of leaves.

For the summer season, the land becomes infested with insects and oppressed by drought. Rain, when it comes, is accompanied by violent lightning that dances among the pines and hammers ancient oaks. The seemingly endless summer begins to soften in October when wild flowers bloom in profusion. Cooler nights forecast the coming of hunting season. Pine straw blankets the forest floor. Briars and blackberry bushes dry and harden to torment dogs and hunters.

Late November rains replenish cypress swamps. Days are alternately brilliant blue or overcast. The streambeds and hardwood bottoms turn gray as the leaves fall to the ground. In contrast, the pines surround them in an array of green. Shadows deepen as the sun travels closer to the southern horizon. Land and season seem in harmony again as the harsh spell of summer is broken.

Sunrise
Chinquapin Plantation
Suwannee County, Florida

Oak Trees and Early Morning Shadows
Popefield Plantation
Calhoun County, Georgia

Road Through Pines
Aucilla Plantation
Thomas County, Georgia

Fire Blackened Pines
Campbell Farms
Dougherty County, Georgia

Controlled Burning
Hooterville Plantation
Terrell County, Georgia

Controlled Burning
Hooterville Plantation
Terrell County, Georgia

Tree Ferns
Greenridge Plantation
Thomas County, Georgia

Magnolia Blossom
Magnolia Plantation
Dougherty County, Georgia

Live Oak Beside Road
Pineland Plantation
Baker County, Georgia

Oak Grove
Pinebloom Plantation
Baker County, Georgia

Sandy Road Through Pines
Ichauway Plantation
Baker County, Georgia

Entry Drive
El Destino Plantation
Jefferson County, Florida

Flint River
Senah Plantation
Lee County, Georgia

Lake Iamonia
Bordered by Plantations
Leon County, Florida

Summer Sky over Water
Ayavalla Plantation
Leon County, Florida

Long Pond
Pinebloom Plantation
Baker County, Georgia

Rex Pond
Pineland Plantation
Baker County, Georgia

Water Hyacinths and Cypress
Lake Iamonia
Leon County, Florida

Cypress Tree
Easter Plantation
Thomas County, Georgia

Lily Pads
Magnolia Plantation
Dougherty County, Georgia

Cypress Tree in Flint River
Riverview Plantation
Mitchell County, Georgia

*Sunset over Pond
Danville Plantation
Sumter County, Georgia*

Quail Shooting

THE YEAR'S FIRST FROST and occasional rains accompany the cooler air of late November. Broom sedge ripens to a golden brown, and the fields emit the earthy smells of a late, southern fall. Bird season is here. Guns are oiled and the dogs are ready. Matched buckskin mules stand waiting, their harnesses festooned with polished brass. The scene is set for quail hunting in the grand plantation style.

It is cold and damp in the gray light of dawn as the men and women gather in the "big house." Breakfast is hurried but plentiful—black coffee and country ham awaken the senses. Outside there is barking in the kennels as the dog handlers make their choices for the day's shooting. "Load up," they command as eager pointers leap into cages mounted on the hunting wagons. Others yelp and surge against the kennel fences, frustrated to be left behind.

After last-minute attention to guns, shells, and gear, the hunters board a mule-drawn wagon. High slung and well sprung to carry its passengers over the bottoms and rough terrain in a semblance of comfort, the wagon creaks forward as a white-coated driver settles the mules with a word and a slap of the reins.

Astride a Tennessee Walking Horse, the dog handler guides the procession to the cover he hopes will reveal a covey of quail. The wagon stops and two dogs are released. They are off, crisscrossing the ground, guided by the handler's signals and their own keen sense of smell. "Look close now, hunt close," he shouts. The pointers' ranging becomes more deliberate.

Suddenly one of the dogs stops still, rigid and intent, tail straight up, forepaw raised in an instinctive gesture. His partner immediately backs the point. "Steady, boy, steady," commands the handler in a low confidential tone. Two hunters dismount, pull double-barreled shotguns from saddle scabbards, load up and approach the concealed birds. In a burst of whirring wings, the birds flush.

Momentarily immobilized by their own anticipation and the intensity of the covey rise, the shooters recover quickly. One member of the party fires first at a low-flying bird that has broken to the left. The bird topples. His swing is too late for a shot at one of the other birds streaming away. His partner waits, then fires her open barrel, downing two birds flying dead away. The surviving birds rise deceptively fast and dip sharply as one last shot proves futile. Out of range within seconds, the birds blur into brown cover and disappear. As the dogs retrieve the day's first birds, a distant popping of guns is heard.

Loading Up
Ashburn Hill Plantation
Colquitt County, Georgia

Matched Mules
Ashburn Hill Plantation
Colquitt County, Georgia

Dogs in Wagon Kennel
Cane Mill Plantation
Dougherty County, Georgia

Hunting Wagon with Retrievers
Mercer Mill Plantation
Worth County, Georgia

Hunting Wagon
Cane Mill Plantation
Dougherty County, Georgia

Unloading in the Field
Gillionville Plantation
Dougherty County, Georgia

Before the Hunt
Borderline Plantation
Thomas County, Georgia

Hunter, Guide, and Dog
Campbell Farms
Dougherty County, Georgia

Hunting Guide Signaling a Point
Ashburn Hill Plantation
Colquitt County, Georgia

Scabbard and Plantation Saddle
Aucilla Plantation
Thomas County, Georgia

On Point
Campbell Farms
Dougherty County, Georgia

Hunter and Dog
Campbell Farms
Dougherty County, Georgia

Hunter and Dog
Three Creeks Farm
Decatur County, Georgia

Quail Hunting
Gillionville Plantation
Dougherty County, Georgia

Holding the Horses
Gillionville Plantation
Dougherty County, Georgia

Covey Rise
Campbell Farms
Dougherty County, Georgia

Retrieving Downed Quail
Idlegrass Plantation
Decatur County, Georgia

Hunting Guide and Dog with Bird
Campbell Farms
Dougherty County, Georgia

Hunting Guide with Dogs
Campbell Farms
Dougherty County, Georgia

Hunting Wagon
Gillionville Plantation
Dougherty County, Georgia

Hunting Wagon
Ashburn Hill Plantation
Colquitt County, Georgia

Water Break
Cane Mill Plantation
Dougherty County, Georgia

Loading Up
Cane Mill Plantation
Dougherty County, Georgia

Loading Mules into Horse Trailer
Gillionville Plantation
Dougherty County, Georgia

Heading Back after the Hunt
Ashburn Hill Plantation
Colquitt County, Georgia

Returning Dogs to Kennel
Ashburn Hill Plantation
Colquitt County, Georgia

Cooling Off
Cane Mill Plantation
Dougherty County, Georgia

Mule Harnesses
Ayavalla Plantation
Leon County, Florida

Plantation Saddle on Hitching Post
Dixie Plantation
Grady County, Georgia

Wagon Wheel
Sedgefield Plantation
Thomas County, Georgia

Quail and Shotgun
Cane Mill Plantation
Dougherty County, Georgia

HANK MARGESON

During the past several decades, Hank Margeson has been working on projects that interpret and record the human qualities and sense of place intrinsic in the diverse cultures of the American South. His documentary approach reflects the influence of photographers August Sander and Walker Evans. Photographs from his plantation work and from other documentary projects are included in numerous museum and corporate collections and have been exhibited in Rome and New York and in other invitational, juried, and one-person exhibitions. Mr. Margeson lives near Dahlonega, Georgia, with his wife and daughter. He teaches photography at North Georgia College.

JOSEPH KITCHENS

Joseph Kitchens received his Ph.D. in history at the University of Georgia and was a professor at Georgia Southwestern in Americus, Georgia, for fifteen years. His interest in historical preservation and architecture led to his 1983 appointment as the first director of Pebble Hill Plantation in Thomasville, Georgia. This museum houses one of the south's finest collections of sporting art and has a century-long history as a quail plantation. With his wife and children, Dr. Kitchens lives on this three-thousand-acre estate. He enjoys sculpting, writing, and shotgunning.